SEQUOYAH REGIONAL LIBRARY
3 8749 0015 4996 1

ASK ISAAC ASIMOV

WHAT HAPPENS WHEN I FLUSH THE TOILET?

BY ISAAC ASIMOV AND ELIZABETH KAPLAN

Gareth Stevens Publishing
MILWAUKEE

Library of Congress Cataloging-in-Publication Data

Asimov, Isaac, 1920-
What happens when I flush the toilet? / by Isaac Asimov and Elizabeth Kaplan.
p. cm. — (Ask Isaac Asimov)
Includes bibliographical references and index.
Summary: Briefly describes how toilets are used to get rid of waste and how sewage is treated.
ISBN 0-8368-0801-0
1. Sewage disposal—Juvenile literature. 2. Sewage—Juvenile literature. [1. Toilets. 2. Sewage disposal.] I. Kaplan, Elizabeth, 1956- . II. Title. III. Series: Asimov, Isaac, 1920- Ask Isaac Asimov.
TD741.A84 1993
628.3—dc20 92-32553

Edited, designed, and produced by
Gareth Stevens Publishing
1555 North RiverCenter Drive, Suite 201
Milwaukee, Wisconsin 53212, USA

The book designer wishes to thank her parents for allowing their bathroom to be photographed and Gilbert Fuiten from Fuiten's Septic Service for his kind cooperation.

Picture Credits
pp. 2-3, Paul Miller/Advertising Art Studios, 1992; pp. 4-5, © Nigel Blythe/Robert Harding Picture Library; pp. 6-7, © H. Armstrong Roberts; pp. 8-9, Paul Miller/Advertising Art Studios, 1992; pp. 10-11, Paul Miller/Advertising Art Studios, 1992; pp. 12-13, Paul Miller/Advertising Art Studios, 1992; pp. 14-15, © Ken Novak, 1992; pp. 16-17, © Camerique/H. Armstrong Roberts; pp. 18-19, Courtesy of McDonnell Douglas Corporation; p. 18 (inset), © Gareth Stevens, Inc., 1992; pp. 20-21, Paul Miller/Advertising Art Studios, 1992; pp. 22-23, © George Hunter/H. Armstrong Roberts; p. 24, © George Hunter/H. Armstrong Roberts

Cover photograph, © Ken Novak, 1992: A private bathroom in a country home in the American Midwest has modern conveniences along with rustic charm.

Series editor: Valerie Weber
Editors: Barbara J. Behm and Patricia Lantier-Sampon
Series designer: Sabine Beaupré
Book designer: Kristi Ludwig
Picture researcher: Diane Laska

Printed in the United States of America

1 2 3 4 5 6 7 8 9 98 97 96 95 94 93

Contents

Words that appear in the glossary are printed in **boldface** type the first time they occur in the text.

Modern-Day Wonders

Pick up your telephone and have a private conversation with someone halfway around the world. Turn on your computer and play games or write a letter. These are only a few of the many wonders of **technology**.

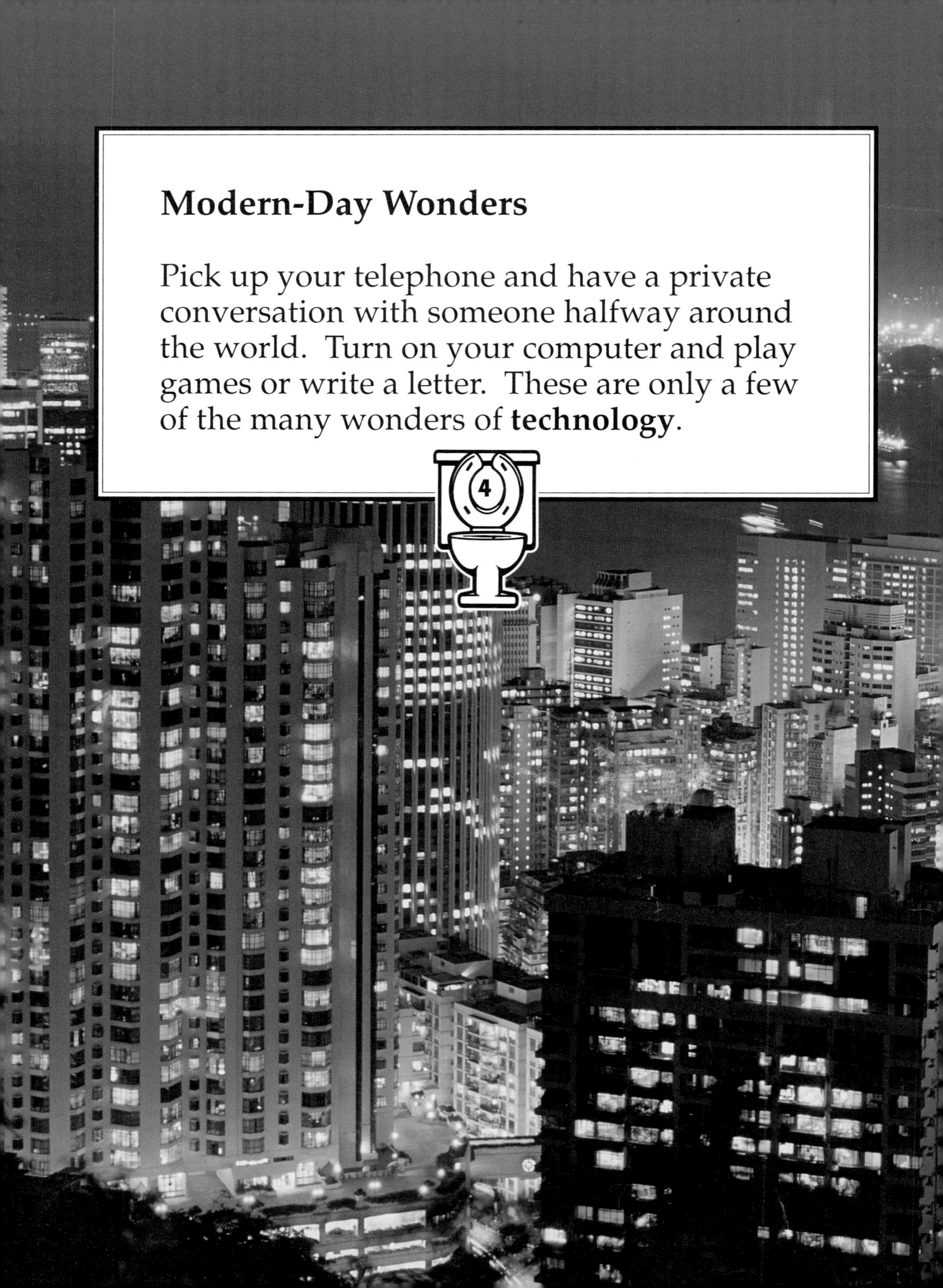

Flush toilets are another important part of our modern world. Most link into a water and waste disposal system that is large and complex. What happens when you flush the toilet? Let's find out.

Wells, Wastes, and Wonderful Water

People can't live without water to drink. In prehistoric times, people gathered water daily from rivers, lakes, or springs. But as villages grew into towns and towns grew into cities, water supplies were no longer left up to nature. In some cities, people dug wells. In others, pipes were built to carry water to homes. Before long, someone decided to use the running water to flush away wastes. The first known toilets were installed in the city of Mohenjo-Daro, in India, about 5,000 years ago.

Far from a Stone Seat

The toilet in Mohenjo-Daro was just a stone seat over a chamber that drained to the street. Today's toilet has a few more parts. The bowl is the part you sit over. The tank holds water for flushing wastes from the bowl. When you push the handle, a lever lifts a flapper in the tank. Then, water rushes from the tank into the bowl and down the drainpipe. This flushes the wastes away. As the water level in the tank falls, a float resting on top of the water drops, opening a part called the inlet valve. Water flows in to refill the tank.

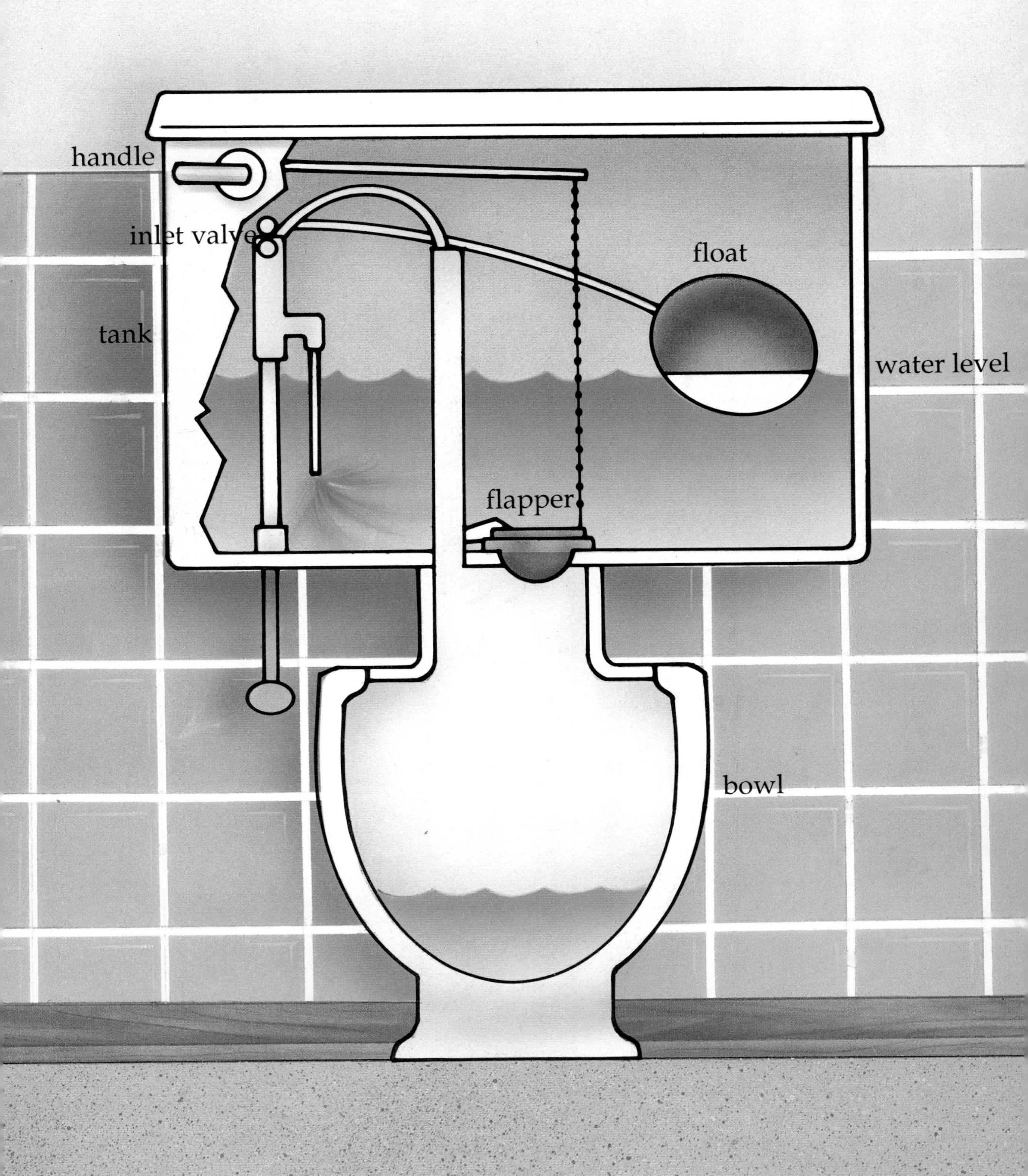
handle
inlet valve
float
tank
water level
flapper
bowl

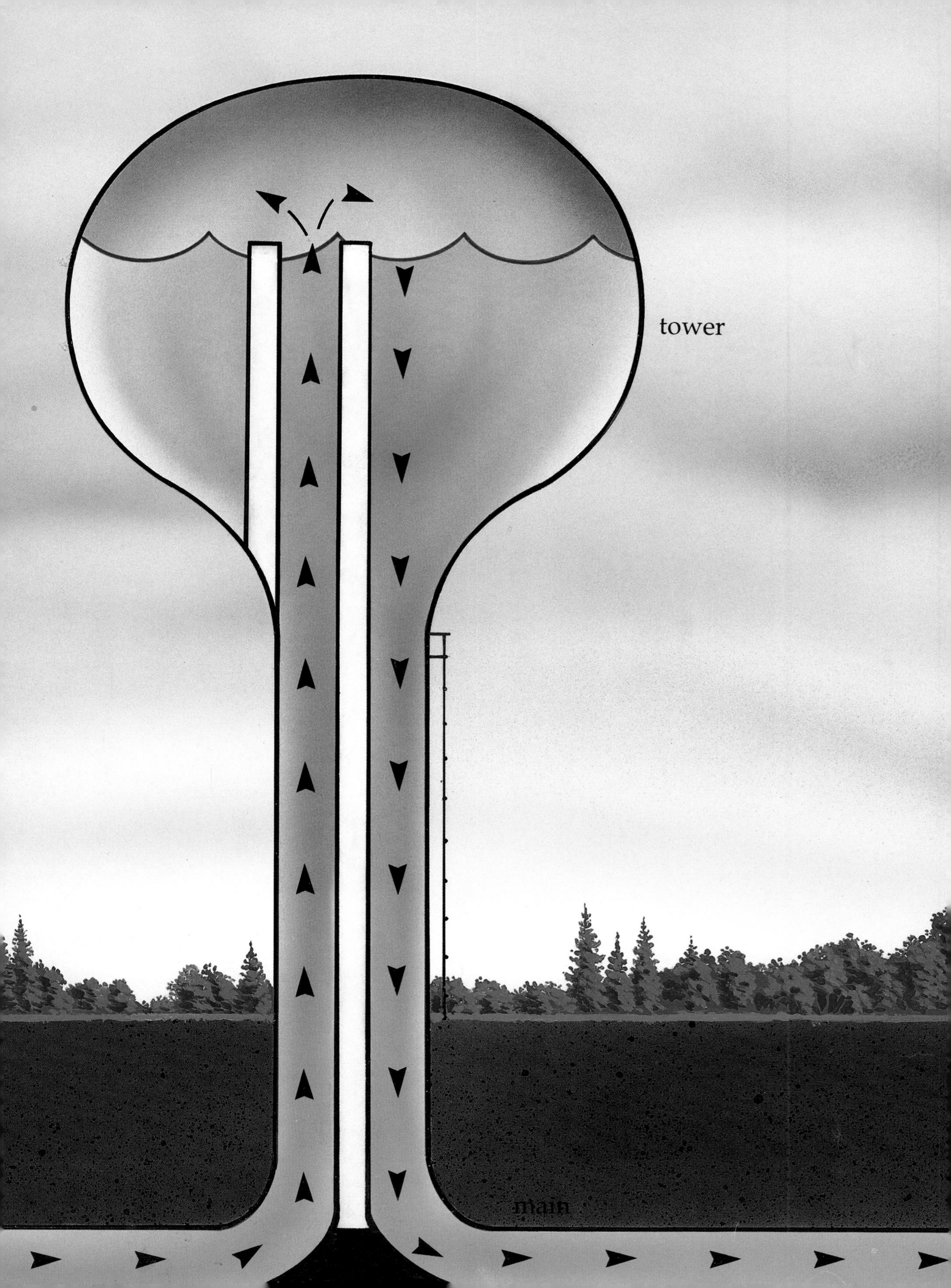
tower
main

Water: Direct to You

Without running water, the flush toilet won't flush. Houses in the countryside often get their water from private wells. Houses in the city are usually connected to a central water system. The water may come from lakes, rivers, or **aquifers**. It is pumped to towers high above the city. Pipes called **mains** carry the water, under great pressure, away from the tower. An intake pipe leads from the main to your house. Other pipes branch off the intake, bringing water to your toilet and faucets.

intake pipe

Rushing Water, Venting Gases

You flush. Whoosh! It's gone. Pipes leading from the toilet — and from the sinks and bathtubs in your home — whisk wastes away.

But dangerous and foul-smelling gases can build up in these pipes. So all waste pipes connect to the **stack**, a large, open pipe that passes through the roof. The stack vents gases outside your house. Clean water, which fills the bowl after you flush, also keeps gases from bubbling back up through the toilet.

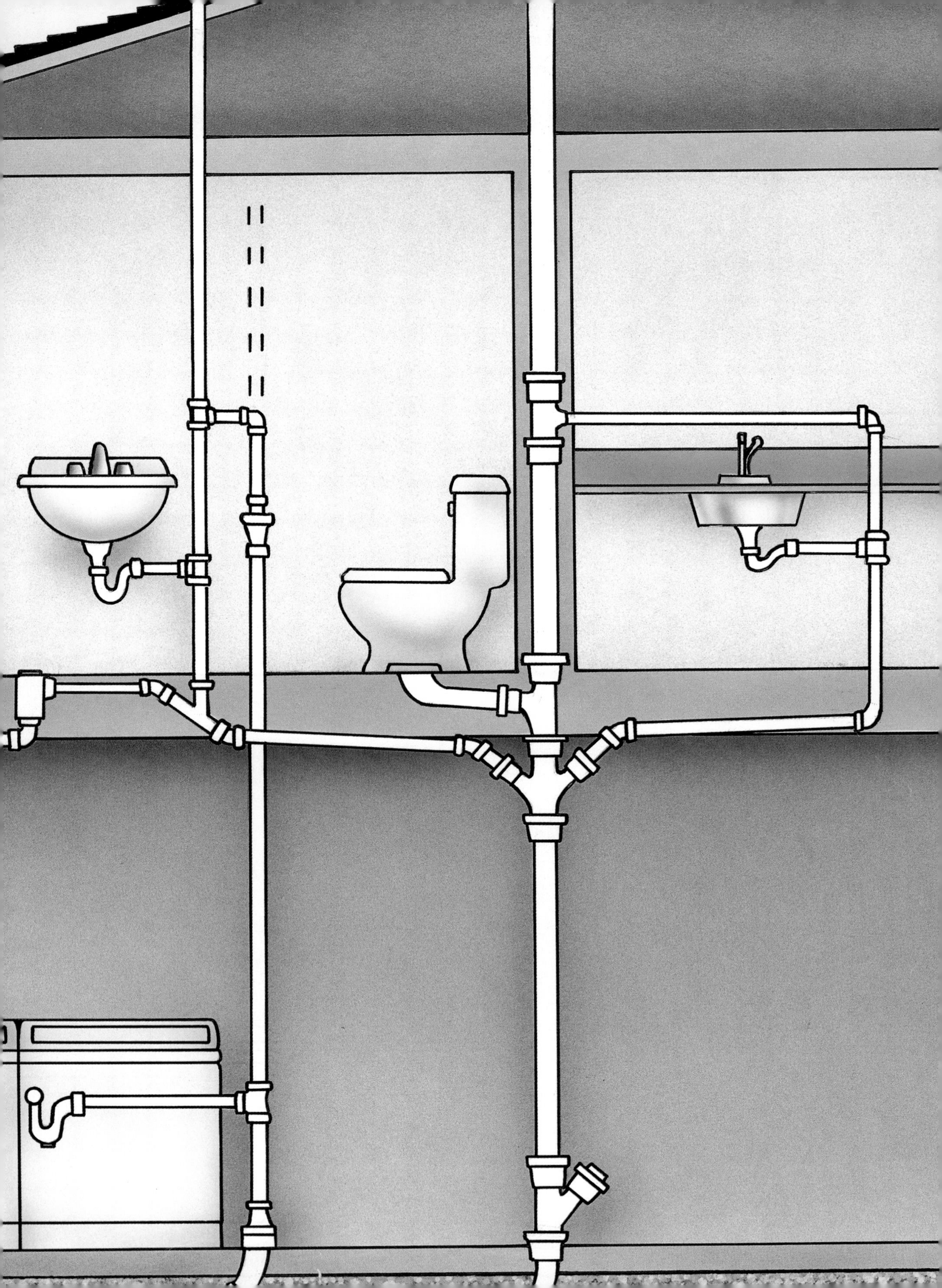

WIS SANITARY LISCENSEE 1424
1000 GAL

Cleaning Wastes — Country Style

Waste from your toilet, called **sewage,** eventually reaches one of Earth's natural sources of water, such as rivers, lakes, or seas. In its raw form, sewage contains disease-causing germs. If it is not treated, sewage can cause serious health problems.

Houses in the country have their own systems called **septic tanks** to clean sewage. Solid wastes sink to the bottom of these underground tanks. There, **bacteria** break down the wastes. About every two years, trucks pump out the wastes from the tanks and take them to be processed. In the city, waste pipes from houses carry wastes to a sewage treatment plant.

The Sewage Treatment Plant

At a sewage treatment plant, wastewater is filtered and cleaned to remove solid particles and disease-causing germs.

Some sewage treatment plants include an extra step to make the water even cleaner. They may use a microscopic screen to remove invisible dirt. They may treat the water with radiation to kill even more germs.

The Toilet on the Plane

If you've ever used a toilet on a plane, bus, or train, you probably noticed that the flushing liquid isn't water. It's a blue, sweet-smelling chemical that prevents wastes from breaking down. The chemically bathed wastes are stored in a holding tank under the vehicle. At stopping places, the tank is emptied. Then, the wastes are taken to a sewage treatment plant, where they are processed like other sewage.

McDONNELL DOUGLAS

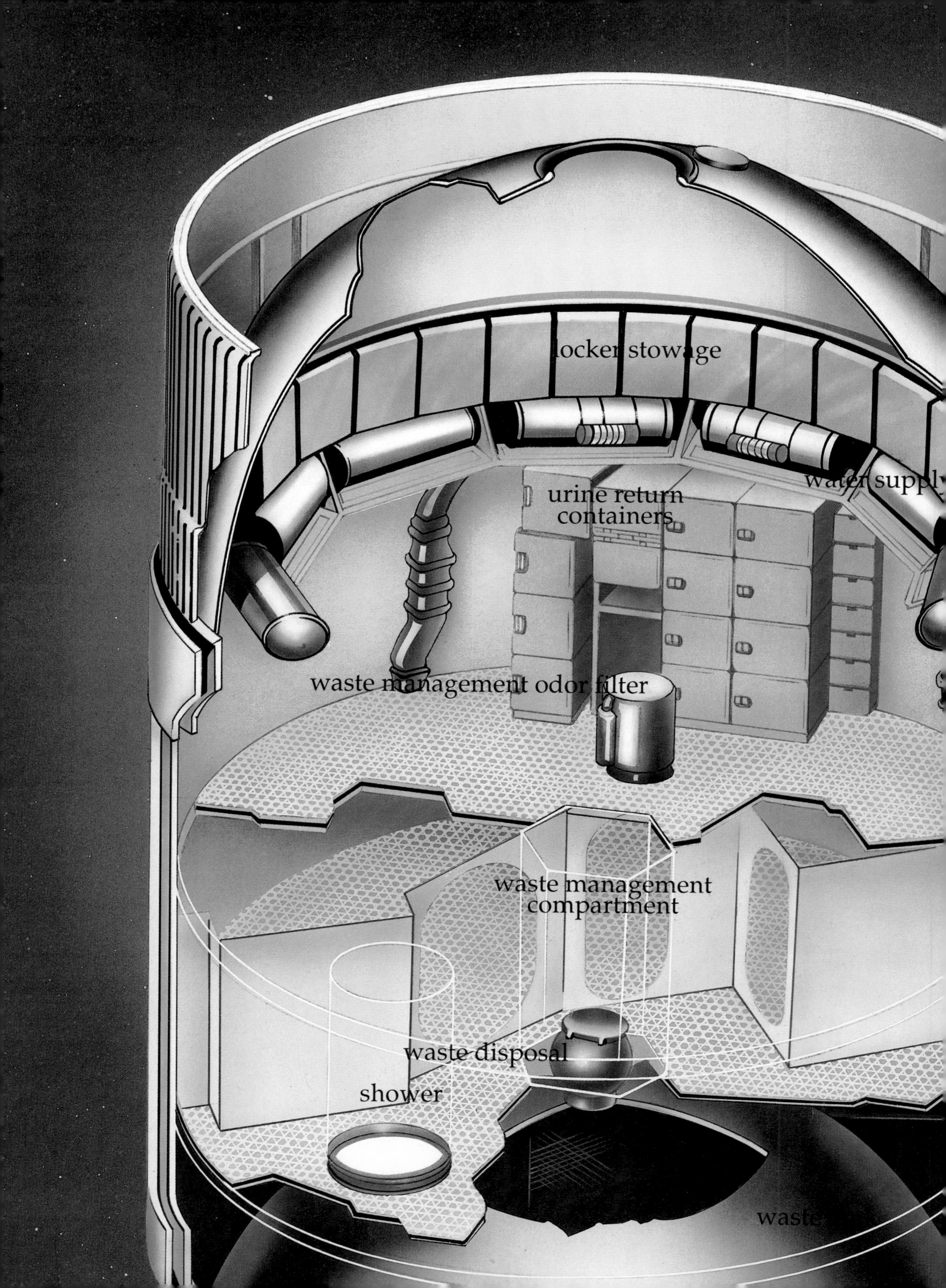

locker stowage
water suppl
urine return
containers
waste management odor filter
waste management
compartment
waste disposal
shower
waste

Wastes in Space

There are no toilets in space. So astronauts need other ways to take care of bodily wastes. Spacecraft have special containers for collecting, drying, and storing these wastes. They aren't processed until the spacecraft returns to Earth.

In the future, people may live in space. Space stations won't have room to store wastes until they can be sent back to Earth. Instead, factories on the station will turn wastes into fertilizer for farms.

Problems with Flushing

Toilets flush wastes away in seconds. But they don't get rid of problems our wastes cause. Flushing requires many gallons of water. But water is becoming ever more scarce, and waste treatment uses a lot of energy. And even after they are treated, wastes still **pollute** beautiful waterways. **Recycling** wastes, as on space stations, may also be best for our Earth.

More Books to Read

I Can Be a Plumber by Dee Lillegard and Wayne Stoker (Childrens Press)
Magic School Bus at the Water Works by Joanna Cole (Scholastic)
The Trip of a Drip by Vicki Cobb (Little, Brown)

Places to Write

The best place to get information about sewage and sewage treatment is probably your own city, or if you live in the country, the nearest large town. Look in the government directory (the blue pages) of your telephone book under "sanitary district," "sewers," or "sewage." Then, you can write the people who work at your local sewage treatment plant to get more information. For more information about toilets and plumbing, you can write to the organization listed below. Be sure to tell them exactly what you want to know. Give them your full name and address so they can write back to you.

Plumbing Information Bureau
303 East Wacker Drive
Chicago, Illinois 60606

Glossary

aquifer (AK-wuh-fuhr): an underground river or lake.

bacteria (bak-TEER-ee-uh): tiny, almost invisible creatures that live in the soil, water, and air. Some bacteria help break down human wastes.

main (mane): a pipe that carries water from a water tower or tanks under the streets of a town.

pollute (puh-LOOT): to add harmful dust, liquids, or gases to the environment. Sewage causes water pollution.

recycling (ree-SIE-kling): cleaning and processing waste materials to turn them into useful products.

septic tank (SEP-tick tank): an underground tank where sewage from a single house collects. Bacteria in the tank break down harmful substances in the sewage.

sewage (SOO-ihj): wastewater, including the liquids and solids, that flows through sewers.

stack (stak): an open pipe that leads through the roof of a house or building to vent gases in the waste pipes.

technology (tehk-NAH-luh-jee): the use of scientific principles to produce things that are useful to people.

Index